Science of Electricity

Volume 5

Solar Power Technologies Explained Simply

by Mark Fennell

© 2012

This book is part of the
Energy Technologies Explained Simply™ Series

Other Books in the Energy Technology Series

Renewable Energy Books
Introduction to Electrical Power
Hydropower Technologies Explained Simply
Solar Power Basic Concepts
Practical Considerations of Solar Power
Advanced Solar Cell Technologies
Wind Power Technology Explained Simply

Coal Power Books
Formation and Mining of Coal
Clean Coal Technologies
Mercury and Coal Power

Nuclear Power Books
Nuclear Power Meltdowns and Explosions
Health Hazards of Radioactive Decay
Radiation Measurements
Processes of Radioactive Decay and Storage of Nuclear Waste

Natural Gas Books
Natural Gas Basics
Extracting and Refining Natural Gas (includes Fracking)
Transportation, Storage, and Use of Natural Gas

Power Line and Grid Books
Introduction to the Transmission of Electrical Power
Power Lines
Underground Cables
Utility Operations and Quality Control
Power Grids Explained Simply

About the Book

This book discusses everything you need to know about solar power. All of your questions about solar power will be answered in this one resource.

- If you want to install a solar array in your home, this book will provide all the information you need.

- If you are a policy-maker, then this book will provide the overview you need to understand all the factors.

- Perhaps you are considering a career in solar power technologies. If so, this book will help you prepare for your future by explaining the most important concepts behind solar power technologies.

There are three major parts to this book:
1. Overview of Solar Power Technologies
2. Practical Considerations when Installing Solar Cells, and
3. Advanced Solar Cell Technologies.

The first part provides an overview of solar cell technologies. Topics include the basic process of solar cells, processes which limit the efficiency of solar cells, and key terminology. This part of the book also provides an overview of practical issues to consider when using solar power.

The second part of the book discusses all of the most important practical considerations when installing a solar power device. Topics include: Electronic Design Choices, Sizing the Array, Orientation, Tilt Angle, and Maintenance. These topics are discussed in detail throughout three chapters.

The third major part of this book discusses the most important advanced solar cell technologies. Topics include the reasons for low efficiency in a solar cell, and the basic methods for improving efficiency. Subsequent topics include: thin films, anti-reflective coatings, surface texturing, minority carriers, solar concentrators, and stacking.

This book also provides an extensive data tables, including:
- tilt angle guidelines based on region and season
- energy required to excite the materials of most solar cells
- and several other data tables and illustrations

Solar Power Technologies Explained Simply is a practical and easy-to-read resource for anyone interested in Solar Power.

About the Energy Technology Series

Purpose of this series

The books in the *Energy Technologies* series are designed to educate citizens, students, and legislators on all aspects of energy technologies. The first books in the series focus on electrical power.

The books discuss many energy technologies, including: generators, turbines, power plants, power lines, and grids. The technologies for each type of power source (hydro, wind, solar, coal, nuclear, and natural gas) are discussed in detail. The books also discuss efficiency, safety, reliability, and health concerns for each energy technology.

The ultimate goal of the series is to enable the people to make informed decisions on practical energy questions. The secondary goal is to serve as introductory guides for students embarking on careers with energy technologies.

Taken altogether, the books in the series answer any question you are likely to have, such as:

- How can we increase the efficiency of solar cells?
- How do I select the size my solar array?
- What do I need to know when installing a wind turbine?
- How effective are the clean coal technologies?
- How can we prevent grid failures?
- Do power lines cause cancer?
- and many other energy technology questions...

Science of Electricity in Perspective

The subject of electrical power is of great importance to our communities, but is rarely taught. Public debate is frequent and passionate, but with too little understanding of the actual science. At best, an informed citizen knows only a few pieces. At worst, as it is for a great number of citizens, electricity is magic and myths are believed as scientific truth. It does not have to be that way. Any citizen, regardless of background, can know the technologies behind all aspects of electricity.

The books in this series solve that problem. These books educate the general public in all aspects of electrical power. Any person, regardless of background, can easily find the answer to his energy question in one of these books.

Specific Goals

There are numerous technologies described in these books. Yet for each technology I sought out the answers to the following questions:

1. How does the technology work?
2. What are the advantages and disadvantages?
3. What is the efficiency? How can the efficiency be improved?
4. What is the environmental impact? How can it be improved?
5. What are the safety hazards, and how can they be reduced?
6. What are the most important practical tips?
7. What facts comprise the most important data?

Technical Discussions Explained Simply

The books in this series explain the principles of electricity as simply as possible, using ordinary English (no engineering jargon), and highlighting the most important points of each technology. Main concepts and facts are emphasized with the use of lists, tables, diagrams, and summaries.

I do not expect any reader to have a background in science, yet I offer enough facts and details so that the reader can have an accurate understanding of all related technologies. I provide enough technical details and enough data for the reader to make informed decisions.

Conclusion

For all the reasons above, I offer this series of books. My goal is to inform you on the basic concepts of all the technologies and all of the issues related to electricity so that you can make realistic decisions.

Remember that there are no perfect solutions, there are only choices. I hope that this series of books will assist you in making those choices for your community.

Mark Fennell

Accuracy and Technical Depth

Objectivity

I have tried my best to be as objective as possible. Whereas many other authors of energy books have an agenda, I have no desire to promote one industry over another. I have no desire to promote one technical solution over another. In this endeavor, I have tried to be an objective scientist.

Accuracy of Data and Summaries

I never relied solely on the conclusions of other researchers. Instead, I performed many other tasks to ensure that all conclusions were accurate. I examined primary data whenever possible. I have read the fine print on how research was obtained.

I have also checked the accuracy of the conclusions written by other researchers, most commonly by finding at least three distinct sources for each fact. In addition, I performed my own calculations numerous times to prove (or disprove) conclusions and final values in other reports. It is only after such rigorous investigations that I created data tables and wrote summaries for these books.

Limited Mathematics

The books must also use math from time to time. For example, efficiency is a statement of a specific amount, and therefore the discussion of efficiency requires the use of equations. Other issues such as power loss, health hazards, environmental concerns, and quality control are also statements of amounts and also require calculations. Therefore some equations are necessary to know, even for the non-scientist.

I also provide examples of calculations so that readers can become more comfortable with using the equation themselves.

However, I want to emphasize that I focus on concepts not on the mathematics. I provide equations only when it is necessary for the citizen or student to be familiar with these equations.

M.F.

Table of Contents

1. <u>Photovoltaic Cells: Basic Concepts</u>   11
a. Introduction
b. Advantages of Photovoltaic Cells
c. Basic Process of Photovoltaic Cells
d. Possible Events in a Photovoltaic Cell and Resulting Efficiency
e. Electromagnetic Waves and Photovoltaic Cells
f. Inverters: Converting DC to AC
g. Solar Cell, Panel, Module, Array
h. Overview of Practical Considerations
i. Temperature and Power in Photovoltaic Cells
j. Technical Details of Photovoltaic Cells

2. <u>Electronic Design Choices</u>   19
a. Introduction
b. Material of Semiconductors
c. Joining Cells Electronically
d. Inverters: Converting DC to AC
e. Stand Alone or Grid-Tied
f. Use Now or Store Later
g. Size of Array – Overview
h. Calculating the Size of Array Needed
i. Batteries and Solar Power

3. <u>Protection and Maintenance</u>   29
a. Material of the protective coating (Encapsulation)
b. Cleaning Schedule and Techniques
c. Physical support for solar cells

4. <u>Orientation to Face the Sun</u>   33

a. Important Points for Array Orientation

b. Latitude and Solar Array

c. Direction: Which Direction is Best

d. Direction: Fixed or Adjustable

e. Tilt Angle: Which Tilt Angle is Best

f. Tilt Angle: Fixed or Adjustable

g. Schedules for Adjusting Tilt Angle

5. <u>Advanced Solar Cell Technologies</u>   41

a. Possible Events in a Solar Cell and Resulting Efficiency

b. Preventing Electron-Hole Pairs From Recombining

c. Using More of Reflected Waves

d. Using Waves That Travel Through the Cell

e. Reducing the Energy Lost as Heat

f. Stacked Solar Cells

g. Thin Films

h. Solar Concentrators Used in Conjunction with Solar Cells

Appendix   53

Bibliography   57

Index   59

5.1
Photovoltaic Cells Basics

Introduction

The word "photovoltaic" means "voltage is produced from light." Photovoltaic cells are commonly called solar cells. I will use both terms throughout this book.

Solar cells are a wonderful use of technology. Although solar cells do not produce much electricity, solar cells can provide a significant contribution. When there are hundreds of buildings in the city with solar power, the total can easily match the power production of any large power plant. The major problem with solar cells is the efficiency. The efficiency of solar cells is very low, at best 20%. However, solar power has many advantages and can be part of a comprehensive electricity program.

Note that some texts will use the term "solar collector." Solar collectors are the same thing as solar cells. However, the better terms are either solar cell or photovoltaic cell. Note also that the term "photovoltaic" is often abbreviated to "PV."

List of topics for this chapter
1. Advantages of Photovoltaic Cells
2. Basic Process of Photovoltaic Cells
3. Possible Events in a Photovoltaic Cell and Resulting Efficiency
4. Electromagnetic Waves and Photovoltaic Cells
5. Inverters: Converting DC to AC
6. Solar Cell, Panel, Module, Array
7. Practical Considerations of Photovoltaic Systems – Overview
8. Temperature and Power in Photovoltaic Cells
9. Technical Details of Photovoltaic Cells

Advantages of Photovoltaic Cells

Photovoltaic cells, commonly known as solar cells, have numerous benefits. Some of the top advantages of photovoltaic cells are:

1. Photovoltaic cells can be put on the top of any roof, supplying electricity to any building.

2. Photovoltaic cells can be used in remote locations, eliminating the need for other sources of energy.

3. Photovoltaic cells can be placed at the point of use, eliminating the need for power lines.

4. Photovoltaic cells are ideal for a thousand purposes such as street lights and communication devices, where the need is only for a small amount of current,.

5. The supply of solar energy will never be used up.

6. Solar energy is accessible to many people.

Basic Process of Photovoltaic Cells (Fig 5.2)

The technical term for a solar cell is a photovoltaic cell. This term literally means that voltage is created from light. When the sun hits an object electrons may become excited. However, most of the time these electrons just return to normal state and are of little use. The solar cell is designed so that those excited electrons are actually turned into a current.

A photovoltaic cell is two semiconductors that are joined together. One of the semiconductors has more electrons than normal, the other semiconductor has fewer electrons than normal. Both semiconductors are necessary in order to make a photovoltaic cell actually work properly.

The first semiconductor produces an exited electron. The material of this first semiconductor is responsive to light. When light hits the first semiconductor, the electrons become excited. This first semiconductor is important, but this semiconductor is only half the story. Without the second semiconductor, the electrons of the first semiconductor might return to normal position, or move in haphazard directions. The second semiconductor forces the excited electrons to flow in one direction, thus creating a current.

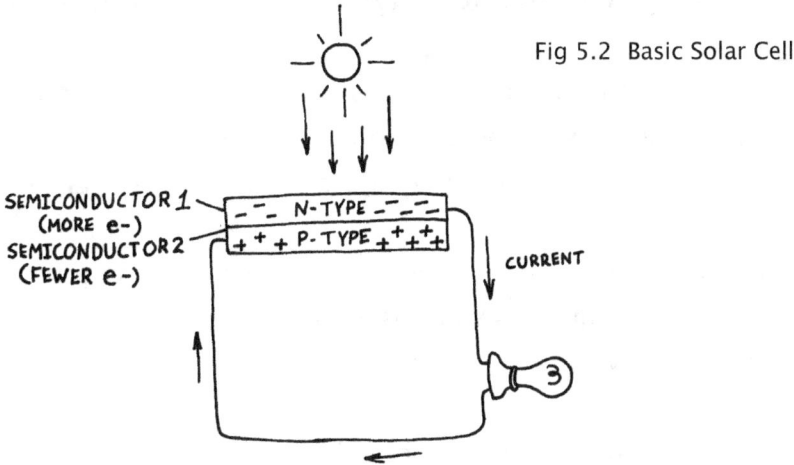

Fig 5.2 Basic Solar Cell

Possible Events in a Solar Cell and Resulting Efficiency

Currently, the efficiency of solar cells is about 15%. This is primarily due to the number of other processes that can happen. In brief, the following are the most common things that can happen when sun hits a solar cell (Figure 5.3):

1. Sun excites electrons. The semiconductors create electron-hole pairs, and therefore current is created. This is what we want.

2. Electron-hole pairs recombine, and thus no current is created.

3. Light waves are reflected.

4. Light waves go through the solar cell, not exciting anything.

5. Excess energy from the light wave is lost as heat.

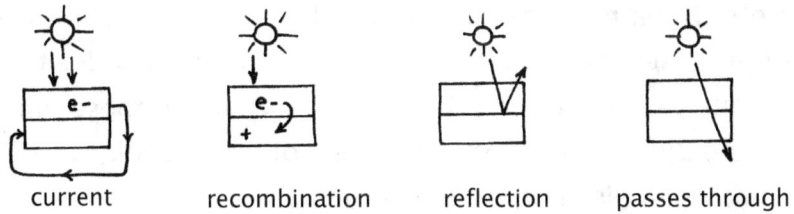

Fig. 5.3 Possible Events in a Solar Cell

Items 2-5 happen most often, 85% of the time. Thus, item 1 (the process we want) occurs only 15% of the time. Methods to improve efficiency are discussed in a later chapter.

Electromagnetic Waves and Photovoltaic Cells

Introduction

The term photovoltaic means that voltage is created from an electromagnetic wave. However, the sun produces many electromagnetic waves. Therefore, designing a photovoltaic cell starts with a two part decision:

1. Decide which wavelengths of electromagnetic energy to work with.

2. Design a material which is responsive to that particular range of electromagnetic waves.

The issues that we must consider when designing the material for photovoltaic cells include:

a. Wavelength and Energy: Shorter waves are more energetic.

b. Amount of waves reaching the earth: The surface of the earth (and hence any solar cell on earth) receives more of some wavelengths than of others.

c. Material response: Each material responds to different wavelengths.

d. Manufacturing: Some materials are easier to make than others, and some materials are cheaper than others.

These issues will be discussed in greater detail in the next chapter.

Units for Wavelength of Electromagnetic Waves

The most common units for wavelengths of electromagnetic waves as used in solar power include: nanometers (nm), micrometers (μm), and meters (m). Note that a micrometer is also called a micron. You may find any of these units used in discussions on solar power. In order to make these various units easier to comprehend, I have arranged these units in a table in the Appendix. For the remainder of this book, I will use nanometers (nm) when discussing solar power.

General Name	Wavelength (nm)
1. Gamma, X–Ray	0 nm – 10 nm
2. UV	10 nm – 400 nm
3. Visible	400 nm – 700 nm
4. Infrared	700 nm – $1 \times 10^{+6}$ nm
5. Microwave, Radio	$> 1 \times 10^{+6}$ nm

Inverters: Converting DC to AC

A photovoltaic cell is unique in the world of electrical generation because this cell produces direct current, DC, rather than alternating current, AC. However, if we want to use electricity in modern buildings and modern appliances, then the DC must be converted into AC. This is the job of the inverter. The inverter for solar cells is a solid state device (electronics). This inverter works by switching the circuit very quickly. The switching of the circuit creates alternating current, and faster switching creates faster frequency.

Solar Cell, Panel, Module, Array (Fig. 5.4)

The various terms for photovoltaic cells include: Solar Cell, Panel, Module, and Array. These terms represent a sequence in size.

1. <u>Solar Cell</u>: A solar cell is one square or one circle. It is one photovoltaic.

2. <u>Panel</u>: A panel consists of several solar cells linked together. A more common term today is module.

3. <u>Module</u>: A module consists of many solar cells linked together. The cells of a module are linked both physically and electronically. A typical module has 36 cells (12 cells x 3 cells), or 48 cells (12 cells x 4 cells).

4. <u>Array</u>: An array consists of several modules connected together. The term "array" applies to the entire set of photovoltaic cells at any particular location.

SOLAR CELL

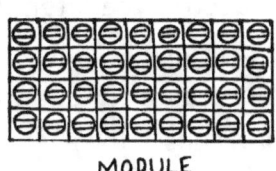
MODULE

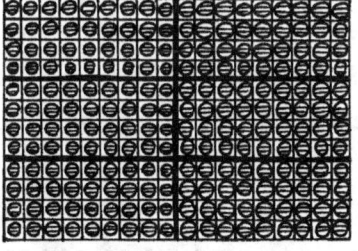
ARRAY

Figure 5.4   Solar Cell, Module, and Array

Thus, a typical rooftop will have an array. This array is made of several modules. The several modules are made of many, many individual solar cells.

Overview of Practical Considerations for PV Systems

When buying and installing an array of photovoltaic cells we should consider the following:

A. Electronic design choices (Chapter 5.2)

Electronic design choices include: material of semiconductors, joining cells electronically, converting DC to AC, batteries, and size of array. These topics will be discussed in chapter 5.2.

B. Protection and Maintenance (Chapter 5.3)

Protection and maintenance choices for solar cells include: 1) material of the protective coating, and 2) cleaning schedule. These topics will be discussed in chapter 5.3.

C. Orientation to face the sun (Chapter 5.4)

Orientation choices for solar cells include: direction to face the array, angle of tilt, fixed or adjustable direction, and fixed or adjustable tilt. These topics will be discussed in chapter 5.4.

Temperature and Power of Solar Cells

The voltage created by a solar cell is inversely affected by temperature. As the temperature of the solar cell increases, fewer Volts will be produced. Since electrical power is related to Volts, we can state the same principle as related to power: When the temperature of the solar cell increases, less power will be produced. When temperature of the solar cell decreases, more power will be produced.

The specific values of power difference as a function of temperature change will depend on the specific solar cells. However, a good rule of thumb for most solar cells is .4% power difference per inverse degree Celsius. This means that for each degree Celsius warmer the solar cell becomes, the amount of power you can get is .4% less. Conversely, for each degree Celsius colder the solar cell becomes, the amount of power you can get is .4% more.

Note that the value of .4% may not seem like much, but when the efficiency of a solar cell is only 15% even in ideal circumstances, a .4% power difference per degree Celsius can make quite a difference.

Technical Details of Photovoltaic Cells

Introduction

In this section we will discuss photovoltaic cells in more technical detail. This section explains some additional concepts regarding the operation of photovoltaic cells. Note that you can skip this section and still use photovoltaic cells effectively. However, some readers will want to know more about the science of photovoltaics and therefore I offer this technical information for their benefit.

Electron-Hole Pairs

Technically, photovoltaic cells work by the creation of "electron-hole" pairs. The energy from the light wave hits the material, excites the electron, and therefore makes the electron able to move. As the electron moves, it leaves a space. The space left behind is a "hole." Both electrons and holes can move around. (If you move electrons around, then the holes must move around to compensate).

n-type, p-type, and doping

Technically, we make two semiconductors: n-type and p-type. In a pure semiconductor material there exists the proper number of electrons as nature intended. However, we can modify the material to suit our needs.

We modify the top semiconductor to have more electrons than the pure material. This semiconductor is called "n-type" (for negative charge). We modify the bottom semiconductor to have fewer electrons than the pure material. This semiconductor is called "p-type" (for positive charge). To get a material with more or fewer electrons, we add specific elements into the mix. Adding these elements is known as doping.

The n-type semiconductor has more electrons than pure material, and therefore it is easier to pull electrons off the material using the light wave than in the natural semiconductor. Once the electron-hole pairs are created by the light wave the electrons and holes can move around. However, if we just had the n-type semiconductor alone, then these electron-hole movements would be haphazard and these electrons would often fall back into place (back into the holes). Therefore, the existence of the p-type semiconductor, with all its built-in holes, forces the electron movements and hole movements into an organized current.

PN Junction and One-Way Current

Semiconductors are very thin, and the electrons in the n-type semiconductor are physically close to the holes in the p-type semiconductor. If left on their own, the electrons would naturally flow down to the holes and combine there. However, that would be of no practical use for us. Therefore, we must build a wall, which will then force the electrons to take a detour. The solution is the PN Junction. The PN Junction separates the p-type semiconductor from the n-type semiconductor (hence the name "PN Junction"). This junction is in essence a wall which prevents electrons and holes recombining in that direction.

A related concept is the diode: you will often hear solar cells referred to as diodes. Recall that a diode is an electronic one-way street, allowing current to flow in only one direction. Therefore, the solar cell is indeed a diode.

Chapter Summary

1. The term "photovoltaic" means that voltage is created from light.

2. A photovoltaic cell has two semiconductors. The first semiconductor responds to a specific wavelength range of light, therefore exciting the electrons. The existence of the second semiconductor makes sure that the excited electrons flow in a current.

3. There are two main issues regarding light and photovoltaic cells:
    a. Some wavelengths of light are more energetic than others
    b. The surface of the earth receives more of some wavelengths than of others.

4. Photovoltaic cells create direct current rather than alternating current. Therefore, in order to use photovoltaic cells we must use an inverter to change from DC to AC.

5. When buying and installing an array of photovoltaic cells we should consider the following:
    a. Electronic design choices
    b. Protection and Maintenance
    c. Orientation to face the sun

5.2
Photovoltaic Cells Practical Considerations: Electronic Design Choices

Introduction

The first subjects we must consider when selecting a photovoltaic system are the electronic design choices. There are several important electronic choices to consider.

List of topics for this chapter
1. Material of Semiconductors
2. Joining Cells Electronically
3. Inverters: Converting DC to AC
4. Stand Alone or Grid-Tied
5. Use Now or Store Later
6. Size of Array – Overview
7. Calculating the Size of Array Needed
8. Batteries and Solar Power

Material of Semiconductor

An important factor in designing a solar cell is the type of materials to use. We noted earlier that the sun produces many different wavelengths. In addition, each material (natural or manufactured) responds to different wavelengths. Therefore we can design materials to respond to the wavelengths that we think are best to work with. The issues that we must consider when designing the material for photovoltaic cells include:

a. Wavelength and Energy: Shorter waves are more energetic.
b. Amount of Waves Reaching the Earth: The surface of the earth (and hence any solar cell on earth) receives more of some wavelengths than of others.
c. Material response: Each photovoltaic material responds to different wavelengths.
d. Manufacturing: Some materials are easier to make than others.

Table A – <u>Wavelengths and Energy</u>: Smaller waves have more energy.

General Name	Wavelength Range	Energy Range
1. Radio, Microwave	larger than $1 \times 10^{+6}$ nm	less than .001 eV
2. Infrared	700 nm to $1 \times 10^{+6}$ nm	.001 eV to 1.75 eV
3. Visible	400 nm to 700 nm	1.75 eV to 3 eV
4. UV	10 nm to 400 nm	3 eV to 125 eV
5. Gamma, X-Ray	0 nm to 10 nm	125 eV to 10^{+7} eV

Table B – <u>Reaching the Earth</u>: Of all the wavelengths that actually hit the surface of the earth, the following wavelengths are the most abundant.

General Name	% of Total
1. Visible (400 nm to 700 nm)	47%
2. Infrared near visible (700 nm to $1 \times 10^{+4}$ nm)	39%
3. Gamma, X-Rays, and UV (0 nm to 400 nm)	7%
4. Larger Infrared, Microwave, Radio (largest waves)	6%

Table C – <u>Material Response</u>: The energy needed to excite electrons of some materials.

Photovoltaic Material	Energy needed (in eV)
1. Lead Selenide (PbSe)	.26 eV
2. Lead telluride (PbTe)	.29 eV
3. Lead Sulfide (PbS)	.37 eV
4. Germanium (Ge)	.70 eV
5. Copper Indium Diselenide (CID)	1.05 eV
6. Silicon (Si) (crystalline)	1.12 eV
7. Indium Phosphide (InP)	1.34 eV
8. Gallium Arsenide (GaAs)	1.42 eV
9. Cadmium Telluride (CdTe)	1.45 eV
10. Cadmium Selenide (CdSe)	1.70 eV
11. Silicon (Si) (amorphous)	1.75 eV
12. Gallium Phosphide (GaP)	2.20 eV
13. Cadmium Sulfide (CdS)	2.40 eV

Other materials for photovoltaics include: aluminum silicon; aluminum antimony; copper sulphide; and copper indium gallium diselenide (CIGS).

Joining Cells Electronically

Solar Cells can be connected in series or in parallel. Solar cells that are connected in series increase the voltage (Figure 5.5). Solar cells that are connected in parallel increase the current (Figure 5.6). Solar cells are usually connected in series in order to get the desired voltage. Solar cells are also connected in combination (connected both in series and in parallel). The particular connections result in different amounts of power.

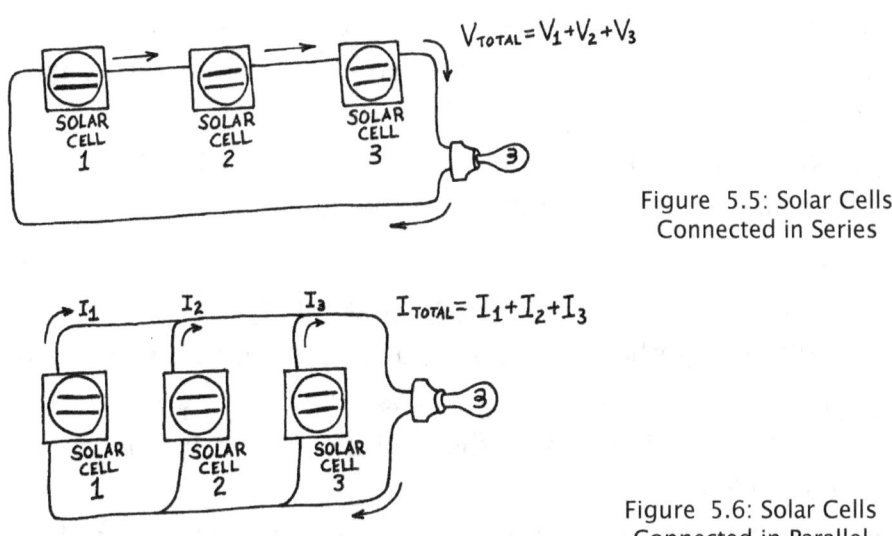

Figure 5.5: Solar Cells Connected in Series

Figure 5.6: Solar Cells Connected in Parallel

Inverters: Converting DC to AC

A photovoltaic cell produces direct current rather than alternating current. Household appliances are designed to operate on alternating current. Therefore, if we want to use household appliances then DC must be converted into AC. This is the job of the inverter. The inverter works by switching an electric circuit very quickly. The switching of the circuit creates alternating current. Faster switching creates faster frequency.

Converting electricity from DC to AC can result in significant loss of power. An inverter can result in power loss of 5% to 40%. This translates into efficiencies from 60% to 95%.

The quality of inverters is very important. The resulting alternating current should have the same quality as electricity coming from the utility company. This means having tight limits on frequency and voltage. Note that if you plan on selling extra solar power to the utility company, then the quality control tolerances must be very strict. (Specific quality control factors are described in a later unit).

It is required by law that inverters must automatically disconnect from the grid when the grid power fails. The reason for this is to provide safety to utility workers. If the solar cell is not disconnected then there is a live system to the grid. Utility workers operating on a line think that this line is dead, but it is not, and the workers can get shocked. Therefore, all inverters must disconnect from the grid system automatically during a power grid failure.

Some inverters make a distinct, audible noise when they operate. The amount of noise an inverter makes depends on the amount of load and the circuit design. As with many other products, quieter devices are available, usually at a higher cost. If the noise cannot be reduced any further by circuitry design, then the inverter should be insulated for sound and placed in remote locations of the building.

Stand Alone or Grid-Tied

When putting together your solar power system you need to decide how independent you wish to be. Most home owners and building managers find that being tied to the grid along with having an array is the most convenient.

Many energy conscious individuals are choosing to go it alone. Theoretically, anyone can be independent from the grid using solar power. However, to make this reality you need two things: 1) a large enough solar array to collect the power and 2) enough batteries to store the power needed for nights and cloudy days.

Going it alone sounds good at first but is not realistic for most people. The reality of needing such a large solar array and the reality of needing so many batteries convinces most solar power users that the safest situation is to remain tied with the grid.

However, small applications can easily be operated by stand-alone solar power. Applications such as street lights, emergency phone boxes, and surveillance cameras do not require much power. Both the solar array and the battery for these applications can be small enough to be very realistic.

The easier option for most people is to continue to be tied to the grid while also using solar power to provide some of the energy needs. You can use the solar power from your array first, then get the remaining power you need from the utility.

You will get some of your power from the utility, yet require less of it. Also you never need to be concerned with storing energy for later.

Use Now or Store Later

Solar energy can be used immediately when created, or the solar energy can be stored for future use. There are generally three options:

- Use solar energy immediately for daily use.
- Store solar energy as it is made; use it later.
- Do both: use solar energy now and store the energy for later.

Combine these options with the choices of grid-tied or stand-alone systems and the choices become numerous. Therefore, the size of your array and the size of your batteries will depend not only on your energy needs but will also depend on how much energy you want to store for later and how independent you wish to be from the utility.

Size of Array – Overview

General

There are two basic approaches to sizing your array:

1. Make it as big as you can, and get whatever power you can
2. Design the array for a particular application

Size of array – Making Array as Big as You Can

This is the sizing approach that most people will use. Doing this will enable you to get as much electricity from solar power as you can, then supplement the rest of your needs from the main utility. The following items are the factors to consider when sizing your array as big as you can:

a. Area available (area on the roof or area in a field)
b. Weight that can be supported by the existing roof
c. Cost of the overall system

Thus, the maximum size of the array to install is an array that: fits within your area on the roof, is within a weight that the roof can support, and is within your budget.

Size of array: Sizing for a Particular Application

The solar array can also be designed for a particular application. This is best for small, remote, stand-alone applications such as: street lighting, communication devices, remote sensors, and refrigeration units.

In this section we will provide an overview of calculating for the size of your array for a particular application. In the next major section we will provide a detailed example of calculating size of the array.

When calculating the size of your array for a specific use these are the factors to consider:

    a. The Energy needed for your application, per day
    b. The Energy of the sun that reaches your location, per day
    c. The efficiency of the solar cell
    d. The efficiency of the inverter (converting DC to AC)

a. The Energy needed for your application, per day

The energy needed for your application can be easily determined. You can figure your energy needs by 1) deciding how many hours your device will be used, then 2) looking at the device for power requirements, and 3) then figuring your total energy needs with that information.

The general calculation for the energy needed for your application, per day will be: # Joules needed = # Watts for device X # seconds device is used per day. This makes the following specific equations:

    • For a device used 24 hours:
    # Joules used per day = # Watts for device x 86,400 seconds

    • For a device used 12 hours:
    # Joules used per day = # Watts for device x 43,200 seconds

b. The Energy of the sun that reaches your location, per day

The energy of the sun that reaches your location depends on many things, including latitude, season, and local geography. Solar Energy is measured in terms of energy per area, per day; usually as kw-hr/$m^2 \cdot$ day. Note that the energy from the sun must be measured at the ground. The energy that reaches the ground is not always as great as the energy that comes from the sun. This is due to various atmospheric phenomena which limits the amount of energy reaching the earth.

Solar energy should be measured every day for at least one year, preferably over several years, in order to get monthly averages and seasonal averages. This data can be obtained from many sources, including local universities and the National Climatic Data Center.

c & d. Efficiency of the solar cell and the efficiency of the inverter

Data for efficiency of the solar cell and the efficiency of the inverter can be provided by the manufacturers.

Calculating the Size of Array Needed

Introduction

Calculating the exact size of the array can be as simple or as sophisticated as you wish. Simpler methods of calculation are easier to do and provide a rough idea of the size needed. However, if you need a more exact calculation then there are formulas which incorporate many types of factors in order to find that optimum size for your array. When deciding what formula to use in this book I have tried to balance between a formula that is simple to use and a certain amount of accuracy. Below is a simple equation for sizing an array, followed by an example.

Size of array =

$$\frac{\text{Energy needed for application, per day}}{[\text{Energy from sun per area} \cdot \text{day}] \times [\text{cell efficiency}] \times [\text{inverter efficiency}]}$$

Example of Sizing Array

a. Energy need for application per day = 7,500 kw–hr per day (24 hr use)

b. Energy from the sun per area, per day= 20,000 kw–hr/m² · day (winter)

c. Efficiency of the solar cell = 15% = .15

d. Efficiency of the inverter = 90% = .90

Therefore, the size of the array needed for this application is:

$$\# \text{ m}^2 = 7{,}500 \text{ kw–hr per day}/[20{,}000 \text{ kw–hr/m}^2 \cdot \text{day} \times (.15 \times .90)]$$
$$= 2.7 \text{ m}^2$$

Batteries and Size of Array

Note that the calculation above takes into consideration the amount of energy needed by a device for both day and night. The array will capture all the energy needed during each day, but we will still need to store that energy in batteries for use each night.

Note that the system must be designed properly taking into account battery and device, as well as day and night. During the day the array will both power the device and provide energy to the battery. During the night, the device will draw its energy from the battery.

In general, the more energy we wish to store the larger the array must be. The simplest way to plan for the array size in conjunction with batteries is to assume that the device is stand-alone as we did for the above calculation. If we figure the energy required for 24 hours and if we use shortest days (which occur during winter) then we will arrive at the uppermost value for the size of the array. That array size will provide all your energy needs. You can then store what you need in batteries.

Batteries and Solar Power

Batteries store energy for later use and are typically used at night or cloudy days when solar power is not available. There are a variety of batteries available for storing solar energy. Lead-acid batteries are the best and most common for use with solar power.

Batteries are essential if the application is to be completely electrically independent. Stand-alone applications are best for smaller power needs such as lighting and communication devices. Note that if you wish to have a solar home that is completely independent of the grid then you will need large solar arrays and lots of batteries.

The size of the batteries you need will depend on how much energy you want to store for later and how independent you wish to be from the utility. Choices of battery type and size should be discussed with a reputable solar energy installer.

Chapter Summary

1. Electronic design considerations for photovoltaic cell systems include the following: choice of materials, joining cells electronically, converting DC to AC, stand alone or grid-tied, size of array, and batteries.

2. The issues regarding material of semiconductors include:
    a. Shorter waves are more energetic.
    b. The surface of the earth receives more of some wavelengths than of others.
    c. Each material responds to different wavelengths.
    d. Some materials are easier to make than others.

3. Solar cells can be connected either in series or in parallel. Solar cells that are connected in series increase the voltage. Solar cells that are connected in parallel increase the current.

4. Solar cells require inverters in order to change DC into AC. Converting electricity from DC to AC can result in significant loss of power: from 5% to 40% power loss.

5. The quality of inverters is very important. The resulting alternating current should have the same quality as electricity coming from the utility company.

6. It is required by law that inverters must automatically disconnect from the grid when the grid power fails. The reason for this is safety to utility workers.

7. There are two basic approaches to sizing your array:
    a. Make it as big as you can
    b. Design the array for a particular application

8. If making the array as big as you can, the factors to consider are:
   a. Area available (area on the roof or area in a field)
   b. Weight that can be supported by the existing roof
   c. Cost of the overall system

9. If sizing the array for a particular application, the factors to consider are:
   a. The energy needed for your application, per day
   b. The energy of the sun that reaches your location, per day
   c. The efficiency of the solar cell
   d. The efficiency of the inverter (converting DC to AC)

10. Calculating the exact size of the array can be as simple or as sophisticated as you wish. One simple formula for sizing an array for a particular application is as follows:

$$\text{Size of array} = \frac{\text{Energy needed for application per day}}{\text{Energy from sun per area, per day} \times \text{overall efficiency}}$$

11. Lead-acid batteries are the best and most common batteries for use with solar power. Types of lead-acid batteries include: lead-antimony, lead-calcium, and lead-antimony/calcium.

5.3
Photovoltaic Cell Practical Considerations: Protection and Maintenance

Introduction

Solar cells need to be protected and cleaned. Debris of all types can come through the air. This debris includes sand, small rocks, leaves, pollen, and smoke. Any debris sitting on a solar cell will limit the amount of light reaching the cell and hence limit the amount of power the cell will produce. Debris can also scratch the solar cells, causing permanent damage.

In addition, consider the following concept: solar cells are most effective in those places where sun shines most of the year. These locations are often dry areas, including deserts and quarries. In these dry areas, dust and small rocks are blown around all the time often hitting the solar cells. Thus, the very places that are best suited for solar power (because of the amount of sun per year) are also the places that get the grittiest debris. Therefore protection and regular maintenance is essential.

List of topics for this chapter
1. Material of the protective coating (Encapsulation)
2. Cleaning Schedule and Techniques
3. Physical support for solar cells

Material of the Protective Coating

All solar cells need a protective coating. This coating must protect the solar cells from being scratched or broken by debris. The coating must also be as clear as possible in order to let the sunlight come through to the cells. Most solar cells are manufactured with this protective coating.

Solar cell manufacturers often refer to the protective coating as "encapsulation." Types of material for encapsulation include: transparent silicone rubber, ethylene vinyl acetate, and tempered glass.

Of these materials, glass is the most scratch resistant. Glass is also similar in composition to the photovoltaic cell (silicon is often a chemical in both). However, glass is the least flexible.

Cleaning Schedule and Techniques

Cleaning schedules depend entirely on the geographic area. The weather in one part of the country will differ dramatically from another part of the country. The types of debris and rate of debris accumulation on solar cells will differ from region to region. In order to really know the proper cleaning schedule we must study cleaning schedules in the community.

Some universities are experimenting with cleaning schedules at this time. In addition, studies of cleaning schedules and cleaning techniques for solar cells must be done at more local colleges and local universities. This is necessary so that each community has relevant data on cleaning schedules for their area.

Physical Support for Solar Array

If we place a solar array on the top of a roof, then we must be able to physically support that array. The solar array itself is relatively light. However, every additional pound is additional weight which the roof must support. A typical solar array weighs 2-4 pounds per square foot (approximately 21-43 pounds per square meter).

The array must also resist the force of storms. Generally, the more durable the array, the heavier it will be. In addition, the array can collect ice and snow, which adds significant weight.

If you choose to use a tracking device then your roof must support that extra weight as well. Materials and designs for tracking devices vary quite a bit. You can buy a tracking device which is made from light material. However, using lighter material will be more vulnerable to storms.

Many people prefer to build array on the ground. The array can be built much sturdier on the ground, and the array can be much larger. Maintenance is also easier.

Chapter Summary

1. Protection and maintenance considerations for photovoltaic cells include the following:
    a. Material of the protective coating
    b. Cleaning schedule and techniques
    c. Physical support for solar cells

2. Solar cells need to be protected because debris can scratch the cells. Solar cells need to be cleaned regularly because sand, leaves, pollen, and smoke will land on the solar cell, and therefore prevent light from entering.

3. Types of protective coatings include: transparent silicone rubber; ethylene vinyl acetate; and tempered glass. Glass is the most scratch resistant but is the least flexible.

4. Cleaning schedules depend entirely on the geographic area.

5. More studies on cleaning schedules for solar cells must be done. These studies must be done at numerous colleges and universities so that every community has the proper data for its own climate.

6. If we place a solar array on the roof, we must have a strong support for the array, and a strong roof. The roof must be able to support
    a. the weight of the support for the array
    b. the weight of the array itself
    c. the weight of wind and snow (where applicable)

5.4
Photovoltaic Cells Practical Considerations: Orientation to Face the Sun

Introduction

The overall goal is to have the sun's rays arrive perpendicular to the solar array. However, due to factors of astronomy the sun changes its position in the sky all the time. Therefore, we need to 1) know where the sun is going to be at any given day and time, then 2) adjust our array accordingly so that the sun's energy arrives perpendicular to the array.

There are three factors that can affect the position of the Earth in relation to the sun: 1) Latitude, 2) Season, and 3) Time of Day. We will use these factors to find the optimum direction and the optimum tilt angle for a solar array at any location.

List of topics for this chapter
1. Important Points for Array Orientation
2. Latitude and Solar Array
3. Direction: Which Direction is Best
4. Direction: Fixed or Adjustable
5. Seasons and Tilt Angle of Solar Array
6. Tilt Angle: Which Tilt Angle is Best
7. Tilt Angle: Fixed or Adjustable
8. Schedules: Possible Schedules for Adjusting Tilt Angle

Most Important Points of Array Orientation

The following are the most important points regarding array orientation.

1. Best Average Direction: Face the array South

2. Best Average Tilt Angle: Use the angle equal to your latitude

3. Overall Goal: Adjust the direction and tilt of the solar array such that the sun's rays arrive perpendicular to the photovoltaic cells.

4. Tilt Angle and Direction Depend on: latitude, season, and time of day

5. Tilt Angle and Latitude: The further North you live, the steeper the average tilt will be.

6. Tilt Angle and Season: Lower your tilt angle 15 degrees from average tilt in the summer. Raise your tilt angle 15 degrees from average tilt in the winter.

7. Direction in Time of Day: Face the array South-East in the morning, and South-West in the afternoon.

Latitude and Solar Array

If we draw the most direct line from the sun to the earth that line would arrive near the equator. Therefore, most of the light coming from the sun arrives near the equator (latitude 0°). Then, every latitude North from the equator receives less and less direct sunlight.

The first practical result of latitude on a solar array is the overall direction to face the array. The sun shines brightest on the equator, and the equator is South of the United States. Therefore, in the United States we face our solar arrays Southward.

The second practical result of latitude on an array is the average tilt angle. The best tilt angle for a solar array is equal to the latitude where you live. Note that the further North you live in the United States the greater the latitude, and therefore you will need to tilt your array by a much greater angle.

Which Direction to Face Array

Which direction should you face the solar arrays? It depends on your goals.

1. If you want the most overall sun for most of the day then face the solar array South. This is because the light from the sun hits the earth mostly near the equator, and the equator is south of the United States.

2. If you want to use the solar power mostly for heating (such as for winter mornings in cold climates) then face the solar array East.

3. If you want to use the solar power mostly for air conditioning (such as for the summer afternoons in southern states) then face the solar array West.

4. In general, it is best to have an adjustable direction, where your array can face Southeast in the morning and Southwest in the afternoon.

Direction: Fixed or Adjustable

We want our solar cells to face the sun as much as possible. Each day, the sun travels from east to west. Furthermore, because the sun is aimed south of us the sun essentially travels southeast to southwest, each day.

The greater flexibility that we have in turning the direction of the array, the more power we can get from the sun any time of day. However, an array which can turn in more directions requires more mechanics, which adds to the overall cost. The direction of an array comes in these basic options:

a. Fixed direction: The array is fixed in one direction permanently. This is the least expensive.

b. Adjustable direction: The direction of arrays can be changed from controls inside a building. The direction of arrays can also be changed automatically, programmed with a clock. However, the programmable array is the most expensive.

Angle of Tilt

Introduction

At what angle should we tilt the solar array? The optimum tilt angle depends on two factors:

1. The Latitude where we live
2. The Season

Latitude and Tilt Angle of Array

The first factor to consider in tilting your array is latitude. The basic rule turns out to be very convenient: Average Tilt Angle = Latitude Where You Live. The reasons for this require some sophisticated geometry and yet the result turns out to be simple and convenient. (If you are interested in the details then you should study geometry as related to astronomy). The end result is Average Tilt Angle = Latitude.

Seasons and Tilt angle (Figure 5.7)

In each of the four seasons, the sun reaches a different height in the sky. Therefore we must tilt our array accordingly. The Earth is actually tilted slightly relative to the sun. The vertical axis of the Earth (the axis of North Pole to South Pole) is tilted 23.5 degrees off from the vertical axis of the sun. This tilt never changes, regardless of the Earth's position. It is the tilt of the Earth, in conjunction with the Earth's position around the sun, which creates the seasons.

In the winter the sun is very low in the sky. At noon the sun never gets very far above the horizon. As the seasons progress, the sun reaches higher and higher positions each noon. At June 21, the sun reaches its highest noon-time position of the year. As summer becomes fall, the sun gets lower. Each day the sun's highest point each day is just a bit lower than the day before. When winter comes again, starting December 21, the sun reaches its lowest noon-time position in the sky.

Because the sun is at different heights each season we must change our tilt angle. If we did not adjust our tilt angle then we would not get exactly perpendicular rays. The actual tilt angle is inversely related to the season. When the sun is lower (winter) we must raise our tilt angle. When the sun is higher (summer) we must lower the tilt angle. (See Figure 5.7) A complete set of data for tilt angles for major cities, with adjustments for seasons, can be found in the Appendix.

Example of tilt angle: City: Austin, Texas Latitude: 30 degrees

Time of Year	Tilt of solar array, from horizontal
Spring:	30 degrees
Summer – lower tilt to:	15 degrees
Fall:	30 degrees
Winter – raise tilt to:	45 degrees

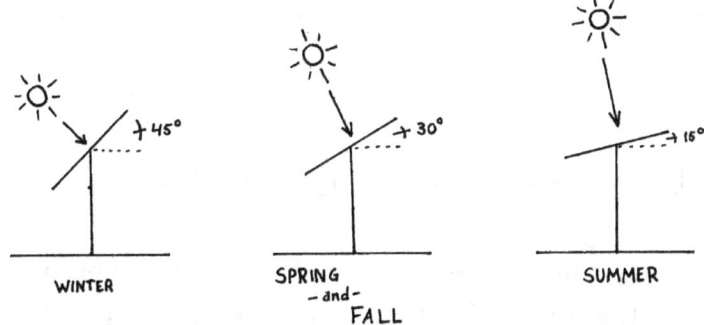

Figure 5.7 Example of Tilt Angle for Each Season

Why not totally flat in the summer?

Note that the tilt is never totally flat in the summer. At the equator we would indeed put the solar cells totally flat in the summer. In America, the slope is low in the summer but it is not flat. For every degree north of the equator we build a solar array, our tilt in summer must one degree higher.

A complete set of data for tilt angles of all major cities, with adjustments for seasons, can be found in the Appendix.

Tilt of Array: Fixed or Adjustable

The tilt angle of the array is just as important as the direction. The greater flexibility we have in adjusting the tilt of the array, the more power we can get from the sun. However, adjusting the tilt requires additional mechanics, which adds to the overall cost. The tilting of an array comes in these basic options:

a. Fixed angle: The array is installed at a particular angle and is fixed permanently.

b. Hand adjustment of tilt: The array is adjusted by hand, usually only a few times each year.

c. Mechanical tilt adjustment: The array can be adjusted from a control inside the building.

d. Mechanical tilt adjustment, with computer program: As with c), the array can be adjusted from a control inside the building. In addition, the computer is programmed for time of day and for calendar day, so that the array always faces the sun at the optimum angle. This system obtains the most power but is also the most expensive.

Schedules for Adjusting Tilt Angle

Introduction

If we want to get the most energy from the sun throughout the year then we must adjust the angle of tilt. We must decide on a schedule for adjusting the tilt angle. There are three basic schedules for tilt angle adjustments: weekly, monthly, and seasonally.

Amount of Tilt

The first decision to make in a tilt angle schedule is how many degrees you want to change your tilt value for a season. An array should be tilted 15 degrees minimum, 23 degrees maximum, throughout each season. Regarding the exact value of seasonal tilt adjustment, different sources say different things. The most common values that I have seen for seasonal tilt changes are 15 degrees, 20 degrees, and 23 degrees.

Many solar array users find that using 15 degrees (changing the tilt 15 degrees each season) is most convenient. Thus, during the middle of summer, when the sun is at its highest, you will have lowered the array 15 degrees less than the average tilt value. During the middle of winter, when the sun is lowest, you will have raised the array in the range of 15 degrees above the average tilt value.

Weekly Schedule

Ideally, you will change the tilt of the array a little bit every week. This allows you to keep the solar array facing the sun directly perpendicular for most of the year. For best results, change the tilt one degree per week.

The reasoning for this is that a season lasts about 13 weeks, and we would like to adjust the tilt 15 degrees throughout a season. Therefore a weekly schedule is approximately 1 degree tilt change per week.

Monthly Schedule

Another convenient schedule is to change the tilt monthly, at five degrees change per month. Doing a five degree change in tilt angle each month, for three months, makes a total change of fifteen degrees over an entire season.

Seasonal Schedule

At the very least, you should change the tilt of the array once each season by the full 15 degrees. Note that the final tilting value for each season will occur during the *start* of that season. The table below gives approximate dates for start, end, and middle of each season. Note that the dates for tilt angles and seasons are based on astronomy, not on the temperature.

For example, June 21 is the start of the summer. It is also known as the longest day of the year. At noon on June 21 the sun will be at its highest point of the year. Similarly, Dec 21 is the start of winter. It is the shortest day of the year. At noon on Dec 21 the sun will be at its lowest point of the year.

Season	Starts	Mid-Season	Ends
Spring	March 20 or 21	May 5–6	June 20 or 21
Summer	June 21 or 22	August 5–6	Sept 21 or 22
Fall	Sept 22 or 23	November 5–7	Dec 20, 21, or 22
Winter	Dec 21, 22, or 23	Feb 2–4	March 19 or 20

Solar Position vs. Temperature

It is important to realize that maximum solar position does not occur on the same date as the maximum local temperature. The two are related but they do not occur at the same time. The solar position comes first; this means that the start of each season comes first. Only later, after the sun has moved on does the temperature of the Earth begin to change.

This is because the Earth has a general insulating quality, and it takes time for temperatures to change. (This is why, for example, Winter begins Dec 21 but February is often the coldest month).

Chapter Summary

1. Optimum orientation of your solar array depends on your location and your goals. However, if you want the most overall sun then face the solar array South.

2. The greater flexibility that we have in turning the direction of the array, the more power we can get from the sun any time of day. However, this greater mobility of the array requires more mechanics, which adds to the overall cost.

3. The optimum tilt angle depends on the latitude and the season. For most of the year, the tilt angle is the same value as the latitude. For summer, the angle is lowered by 15 degrees. For winter, the angle is raised by 15 degrees.

4. A greater flexibility we have in adjusting the tilt of the array results in more power we can get from the sun. However, adjusting the tilt requires mechanics which adds to the overall cost.

5.5
Advanced Solar Cell Technology

Introduction

The efficiency of a typical solar cell is only 15%. If we are to improve the efficiency of a solar cell then we must find a way to engineer around the unwanted processes that occur when sunlight hits a solar cell.

List of topics for this chapter
1. Possible Events in a Solar Cell and Resulting Efficiency
2. Preventing Electron-Hole Pairs From Recombining
3. Using More of Reflected Waves
4. Using Waves That Travel Through the Cell
5. Reducing the Energy Lost as Heat
6. Stacked Solar Cells
7. Thin Films
8. Solar Concentrators Used in Conjunction With Solar Cells

Possible Events in a Solar Cell and Resulting Efficiency

Currently, the efficiency of solar cells is about 15%. This is primarily due to the number of other processes that can happen. The following are the most common things that can happen when sun hits a solar cell:

1. The semiconductors create electron-hole pairs, current is created.
2. Electron-hole pairs recombine, and thus no current is created.
3. Light waves are reflected.
4. Light waves go through the solar cell, not exciting anything.
5. Excess energy from the light wave is lost as heat.

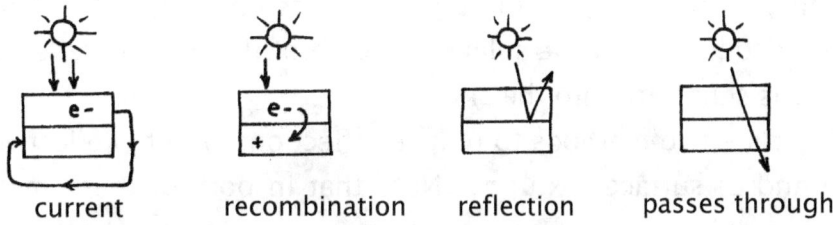

current   recombination   reflection   passes through

Fig. 5.3 Possible Events in a Solar Cell

Items 2-5 happen most often, 85% of the time. Item 1 (the process we want) occurs only 15% of the time. Therefore if we want to increase the efficiency we must find ways to prevent items 2-5 from occurring.

Preventing Electron-Hole Pairs from Recombining

Solar cells work by the creation of "electron-hole" pairs. The energy from the light wave hits the material, excites the electron, and therefore makes the electron able to move. As the electron moves it leaves a space. The space left behind is a hole.

Electrons will always prefer to combine with holes rather than traveling as current. Therefore, we must prevent recombination whenever possible. Electron and holes recombine most often at two locations: 1) at the surface of the semiconductor, and 2) at the contacts (the connections between the cell and the world).

We can prevent recombination at the surface of the semiconductor by coating the external surface with a layer of non-conductive material. This material is usually an oxide or nitride.

We can prevent recombination at any of the contacts by surrounding these contacts with heavily doped material. This doped material can be either positively or negatively doped, depending on where the contact is located. This material is known as a "minority carrier mirror."

Using More of the Reflected Waves

Introduction

Approximately 1/3 of the light which hits the solar cell is reflected. Any energy wave from the sun that is reflected is energy that the solar cell is not able to turn into electricity. Therefore, efficiency can be improved by reducing reflection.

Reflection occurs primarily at the material of first semiconductor. Typically, most light goes through the glass encapsulation, then hits the first semiconductor of the solar cell. Some of that light is reflected off the cell and is sent back into the sky.

There are two methods to reduce reflection: 1) Anti-Reflective Coating (ARC) and 2) surface texturing. Note that in both of these methods the reflection still occurs, however we are containing that reflection within the cell. By containing the reflected wave inside the cell, we give that wave a second chance to be absorbed by the semiconductor.

Anti-Reflective Coating (ARC)

The Anti-Reflective Coating (ARC) is applied either on the surface of the first semiconductor or on the underside of the glass encapsulation. The Anti-Reflective Coating prevents reflected light from passing through the encapsulation. The light is then sent back to the semiconductor to be absorbed. Using Anti-Reflective Coating can we can reduce the amount of escaping waves from 30% down to 10%.

Surface texturing

Surface texturing is the process of making the surface of a solar cell very bumpy. When the light hits the bumpy surface the light is scattered in different directions. This scattering sends the light back to the semiconducting layer and therefore gives the solar cell another chance to absorb the reflected light.

Surface texturing is actually an engineered process, with textures made in particular geometries in order to make the most effective use of reflected waves. When surface texturing is combined with anti-reflective coating, only 1% of the original incoming waves are reflected and still escape.

Using Waves That Travel Through the Cell

Introduction

Some waves will pass right through the cell, not interacting with the material and therefore not exciting anything. The specific waves that pass through the cell depend on the particular semiconductor. There are two main methods to use waves that travel through the cell: 1) use of reflective coating on the bottom and 2) stacking.

Reflective coating on the bottom

The reflective coating on the bottom of a solar cell is much like a mirror. Any light which passes through the cell will hit this reflective coating below and then be reflected back up into the solar cell. This gives the solar cell another chance to absorb the light.

Note that the material for the reflective coating must be specific to the wavelengths we want to reflect. This is because each material will reflect only certain wavelengths.

Stacking

Stacking is essentially putting one solar cell on top of another. Each solar cell is made of a different material, designed specifically to respond to a different wavelength. This means that wavelengths which pass through a cell can now be used by the cell below. By stacking the solar cells, each wavelength from the sun will eventually be turned into electricity.

The efficiency of stacked cells can turn a traditional solar cell of 15% efficiency to become an efficiency of 30% or more.

Reducing the Energy Lost as Heat

Introduction

The energy of the electromagnetic wave must have enough energy to excite the electron in the semiconductor material. However, if the energy of the wave is more than what is needed to create a current, then that extra energy would be simply lost as heat.

Furthermore, the amount of power produced by a solar cell is inversely related to the temperature. Therefore, any additional heat produced will result in less power and a lower efficiency. For these reasons, it is important to reduce heat as a means of improving efficiency.

Stacking and Heat Loss

The best solution for reducing heat loss is through stacking. The first cell in the stack has a material which requires high amounts of energy. Therefore, the highest energy waves are used efficiently, with not much energy left over to become heat. See the section on stacking in this chapter for more details.

Cooling

Cooling any electrical equipment will always improve the efficiency. Regarding solar cells, the wind provides most of the cooling. When possible, it is best to place solar cells in the path of local winds. Also, it is advisable to have enough space on all sides of the array for air to circulate around the cells.

Stacked Solar Cells

In a stacked cell, each solar cell responds to a different wavelength. Wavelengths that normally could not be used by a cell can now be used by the cell below. By stacking the solar cells, each wavelength from the sun will eventually be turned into electricity.

Each layer in the stacked solar cell is made of a different material. The material is selected such that only the most energetic of the waves will create a current. The remaining waves are allowed to pass through the cell to the next layer. The efficiency of stacked cells can turn a traditional solar cell of 15% efficiency to become an efficiency of 30% or more.

Thin Films

Introduction

A "thin film" is essentially a very thin form of the photovoltaic cell. There are many advantages to thin films, most of which are related to economics. The active layer of a thin film generally has a thickness between 1–10 micrometers. For reference, the active layer of traditional solar cells can be as thick as 100 micrometers. Visible waves range in size from .4 to .7 micrometers.

There are several different types of thin films. The most common of these thin films include: amorphous silicon, thin–film silicon, cadmium telluride (CdTe), and copper indium diselenide ($CuInSe_2$). Note that copper indium diselenide is often abbreviated "CID".

Actual Thickness and Optical Thickness

There is a difference between actual thickness and optical thickness. The "actual thickness" is a straight measurement from top to bottom. The "optical thickness" is the total distance that a wave can travel in that layer. Due to a few engineering tricks the optical thickness of the thin film can be much greater than the actual thickness. It is due to these engineering tricks that thin films can be thin and yet still work properly. These two most common methods are: Reflective Coating, and Surface Texturing. We have discussed these methods earlier.

Note that the largest requirements for optical thickness are in the range of 200 micrometers, yet a "thin film" can be 10 micrometers or less.

Substrate

A thin film, by its very nature of being thin, must be attached to something larger. We need something more tangible in order to work with the thin film. This larger piece is known as the "substrate." This substrate must adhere strongly to the thin film. Furthermore, this substrate must also expand and contract along with the thin film during temperature changes. If the substrate and thin film do not expand at the same rate then one or both will crack.

The most common substrate for thin films is glass. In this situation, a module is essentially a large piece of glass, and the thin film solar cell is just a tiny layer on that glass.

What a better material than glass? Both glass and thin films are made of silicon, therefore glass is a perfect thermal match for most thin films. The glass and the thin film will expand and contract with temperature changes at the same rate. Glass is also a familiar material – there are many people who work with glass. These manufacturers know how to shape glass, size it, and modify its properties.

Advantages of Thin Films

There are many advantages to thin films, most of which relate to economics. The primary advantage to thin films is less material, which means that the cost to make each solar cell is much cheaper. Therefore, larger arrays can be more affordable.

The second significant advantage is that modules can be connected during manufacturing. Connecting thin films together into modules can be done when the thin films are being made which thus makes thin film modules much cheaper than traditional modules.

Solar Concentrators as Used with Solar Cells

Introduction

Solar concentrators focus sunlight onto a small area. Therefore, the solar energy is "concentrated" at that location. This small area receives more energy than it would without the concentrators. There are two forms of solar concentrators: lenses and mirrors.

A lens solar concentrator is simply a curved glass. Any curved glass will focus light to a single point. You may have used a magnifying glass during the summer and noted that sometimes the object gets very hot, perhaps hot enough to start a small fire. The magnifying glass is in fact a solar concentrator.

When a mirror is used as a concentrator the mirror is usually shaped as a parabola. The parabolic shape of the mirror reflects each ray of the sun to a specific point. The solar cell is placed at the focus point of the parabolic mirror, thus receiving reflected rays from several directions.

Both lenses and mirrors can be used as solar concentrators on photovoltaic cells. Solar concentrators can be used with photovoltaic cells, increasing the amount of solar energy by 100 times.

Why Solar Concentrators with Solar Cells?

There are two advantages to this method: smaller size and cheaper cost. By using concentrators we can make solar cells smaller. This means that we can build larger arrays of solar cells at a much cheaper cost.

Furthermore, by using a concentrator we can now afford the more efficient solar cells. Higher efficiency solar cells are available, yet these solar cells are more expensive. However, if we use a concentrator then we do not need as large an array. The more efficient solar cells now become more cost-effective to purchase.

Major Technical Concern: Temperature Damage

The major concern with using concentrators on solar cells is that solar cells can become damaged by the increase in heat. The solar concentrator increases the number of all wavelengths coming to a particular point. As we increase the number of wavelengths which will excite our solar cell, we are also increasing the number of other wavelengths which simply heat the cell (without any creation of current.)

Therefore, while we are focusing the energy from the sun onto the solar cells in order to excite more electrons, we must at the same time protect the solar cells from the increase in heat.

Whether a material heats up or not, and by how much, depends on the material's response to each wavelength. Specifically, this depends on which wavelengths excite electrons into current, and which wavelengths excite the molecules into heat without creation of current.

If you are going to use solar concentrators with your photovoltaic cells then you must choose a photovoltaic system that is specifically designed for use with a concentrator.

Engineering a Solar Concentrator System to Reduce Heat

There are a few approaches to reduce heat when using a solar cell:

1. Apply a cooling mechanism in the solar cell. This can be done by providing adequate spacing around the array, creating channels which guide the wind around hot spots, or by installing small fans.

2. Use a version of the "stacking" method above. A stacked cell will use most of the energy for exciting electrons, and less energy wasted as heat.

The Other Disadvantage: Sophisticated Tracking Devices

The other disadvantage of using concentrators with solar cells is that sophisticated tracking devices must be used. Solar concentrators focus the sun's energy to a particular point. However, the sun changes its position throughout the day. Therefore, in order to concentrate the solar energy on the cell we must use a sophisticated tracking device. This tracking device will automatically position the lens relative to the sun and the solar cell at all times.

However, such an automatic device can be expensive, which may not be cost effective. In addition, with a more sophisticated the device there are more possibilities that the device will break.

If your solar array has a fixed orientation and fixed tilt angle, then you can use a solar concentrator with reasonable effectiveness yet without the need of any tracking device.

Summary

1. If we are to improve the efficiency of solar cells then we must reduce the unwanted processes:
    a. recombination of electron-hole pairs
    b. reflected waves going to the sky
    c. waves that travel through the cell
    d. excess energy lost as heat

2. To reduce recombination of electron-hole pairs:
    a. Coat the external surface with a non-conductive material
    b. Surround the contacts with heavily doped material

3. To prevent reflected waves from going to the sky
    a. Use Anti-Reflective Coating (ARC)
    b. Texture the surface

4. To use waves that travel through the cell
    a. Coat the bottom with a reflective coating
    b. Stack layers of solar cells

5. The best method to reduce heat loss in a cell is through stacking.

6. A thin film is essentially a very thin form of the photovoltaic cell, usually between 1-10 micrometers. The thin film is attached to a substrate, usually glass. There are many advantages to thin films, most of which are related to economics. Thin film technology makes solar power much cheaper.

7. In a stacked cell each solar cell responds to a different wavelength. Wavelengths that could not be used by one cell can be used by the cell below. The efficiency of stacked cells can be 30% or more.

8. Concentrators can be used with solar cells. The methods of solar concentration use lenses or mirrors.

9. The first advantage of using concentrators with solar cells is getting up to 100 times more energy focused on each cell. Other advantages are using smaller solar cells and making solar arrays more cost effective.

10. The primary disadvantage of using concentrators with solar cells is the potential damage from overheating. The other disadvantage of concentrators with solar cells is that sophisticated tracking devices must be used.

Conclusion

Many Americans hold passionate views about electrical power, yet few Americans understand all the details behind their passion. Electricity should not be mysterious. The science, the technology, and the data of electrical power can be understood by anyone.

Above all else, we must remember that there are no perfect solutions, there are only choices. Any option can be beneficial, yet each option has its own technical issues to work with. It is up to you and to your community to make those educated decisions. I hope that this book will help guide you in your choices.

M.F.

Appendix

Appendix Items

1. Wavelengths Listed in Common Units
2. Tilt Angle Guidelines for Solar Arrays

Wavelengths listed in Common Units

General Name	Wavelength (nm)	Wavelength (mm)	Wavelength (m)
1. Gamma, X–Ray	0 nm – 10 nm	0 – .01 mm	0 m to 10^{-8} m
2. UV	10 nm – 400 nm	.01 – .40 mm	10^{-8} m to $4x10^{-7}$ m
3. Visible	400 nm – 700 nm	.40 – .70 mm	$4x10^{-7}$ m to $7x10^{-7}$ m
4. Infrared	700 nm – $1x10^{+6}$ nm	.70 – 1,000 mm	$7x10^{-7}$ m to .001 m
5. Microwave, Radio	>$1x10^{+6}$ nm	1,000 – $1x10^{+7}$ mm	.001 m to 100 m

Tilt Angle Guidelines for Solar Arrays
Listed for major cities, and for different times of year
Cities are listed in order of latitude

A. Latitude: 29 degrees: San Antonio, Cape Kennedy

Time of Year	Tilt of solar array, from horizontal
Avg. for all year:	29 degrees
Summer – lower tilt to:	14 degrees
Winter – raise tilt to:	44 degrees

B. Latitude: 30 degrees: Austin, Houston, New Orleans, Tallahassee, Jacksonville

Time of Year	Tilt of solar array, from horizontal
Avg. for all year:	30 degrees
Summer – lower tilt to:	15 degrees
Winter – raise tilt to:	45 degrees

C. Latitude: 31 degrees: El Paso, Waco, Killeen, Hattiesburg, Dothan

Time of Year	Tilt of solar array, from horizontal
Avg. for all year:	31 degrees
Summer – lower tilt to:	16 degrees
Winter – raise tilt to:	46 degrees

D. Latitude: 32 degrees: San Diego, Tucson, Abilene, Dallas, Savannah

Time of Year	Tilt of solar array, from horizontal
Avg. for all year:	32 degrees
Summer – lower tilt to:	17 degrees
Winter – raise tilt to:	47 degrees

E. Latitude: 33 degrees: Long Beach, Palm Springs, Phoenix, Lubbock, Atlanta

Time of Year	Tilt of solar array, from horizontal
Avg. for all year:	33 degrees
Summer – lower tilt to:	18 degrees
Winter – raise tilt to:	48 degrees

F. Latitude: 34 degrees: Los Angeles, Santa Barbara, Wichita Falls, Little Rock

Time of Year	Tilt of solar array, from horizontal
Avg. for all year:	34 degrees
Summer – lower tilt to:	19 degrees
Winter – raise tilt to:	49 degrees

G. Latitude: 35 degrees: Bakersfield, Albuquerque, Amarillo, Ok. City, Memphis

Time of Year	Tilt of solar array, from horizontal
Avg. for all year:	35 degrees
Summer – lower tilt to:	20 degrees
Winter – raise tilt to:	50 degrees

H. Latitude: 36 degrees: Fresno, Las Vegas, Tulsa, Nashville, Raleigh–Durham

Time of Year	Tilt of solar array, from horizontal
Avg. for all year:	36 degrees
Summer – lower tilt to:	21 degrees
Winter – raise tilt to:	51 degrees

I. Latitude: 37 degrees: San Jose, Stockton, Wichita, Richmond (VA)

Time of Year	Tilt of solar array, from horizontal
Avg. for all year:	37 degrees
Summer – lower tilt to:	22 degrees
Winter – raise tilt to:	52 degrees

J. Latitude: 38 degrees: Napa, Sacramento, Colorado Springs, St. Louis

Time of Year	Tilt of solar array, from horizontal
Avg. for all year:	38 degrees
Summer – lower tilt to:	23 degrees
Winter – raise tilt to:	53 degrees

K. Latitude: 39 degrees: Denver, Kansas City, Cincinnati, Philadelphia

Time of Year	Tilt of solar array, from horizontal
Avg. for all year:	39 degrees
Summer – lower tilt to:	24 degrees
Winter – raise tilt to:	54 degrees

L. Latitude: 40 degrees: Salt Lake City, Columbus, Pittsburgh, New York City

Time of Year	Tilt of solar array, from horizontal
Avg. for all year:	40 degrees
Summer – lower tilt to:	25 degrees
Winter – raise tilt to:	55 degrees

M. Latitude: 41 degrees: Redding, Omaha, Des Moines, Chicago, Cleveland

Time of Year	Tilt of solar array, from horizontal
Avg. for all year:	41 degrees
Summer – lower tilt to:	26 degrees
Winter – raise tilt to:	56 degrees

N. Latitude: 42 degrees: Klamath Falls, Sioux City, Detroit, Albany, Boston

Time of Year	Tilt of solar array, from horizontal
Avg. for all year:	42 degrees
Summer – lower tilt to:	27 degrees
Winter – raise tilt to:	57 degrees

O. Latitude: 43 degrees: Boise, Madison, Milwaukee, Syracuse, Concord

Time of Year	Tilt of solar array, from horizontal
Avg. for all year:	43 degrees
Summer – lower tilt to:	28 degrees
Winter – raise tilt to:	58 degrees

P. Latitude: 44 degrees: Eugene, Yellowstone, Pierre, Green Bay, Montpelier

Time of Year	Tilt of solar array, from horizontal
Avg. for all year:	44 degrees
Summer – lower tilt to:	29 degrees
Winter – raise tilt to:	59 degrees

Q. Latitude: 45 degrees: Portland (Oregon), Billings, Minneapolis/St. Paul

Time of Year	Tilt of solar array, from horizontal
Avg. for all year:	45 degrees
Summer – lower tilt to:	30 degrees
Winter – raise tilt to:	60 degrees

R. Latitude: 46 degrees: Astoria, Walla Walla, Helena, Bismarck, Duluth

Time of Year	Tilt of solar array, from horizontal
Avg. for all year:	46 degrees
Summer – lower tilt to:	31 degrees
Winter – raise tilt to:	61 degrees

S. Latitude: 47 degrees: Olympia, Seattle, Spokane, Great Falls, Fargo

Time of Year	Tilt of solar array, from horizontal
Avg. for all year:	47 degrees
Summer – lower tilt to:	32 degrees
Winter – raise tilt to:	62 degrees

Bibliography

Solar Power

1. Energy for Man: From Windmills to Nuclear Power, by Hans Thirring, 1958. Publisher: Indiana University Press.
2. Energy Resources, by Andrew Simon, 1975. Publisher: Pergamon Press, Inc
3. Nontechnical Guide to Energy Resources, by Ben Ebenhack, 1995. Publisher: PennWell Publishing Company
4. Electric Power Generation: A Nontechnical Guide, by Barnett and Bjornsgaard, 2000. Publisher: PennWell Publishing Company
5. Power Surge: Guide to the Coming Energy Revolution, by Flavin and Lenssen, 1994. Publisher: W.W. Norton & Company
6. Energy: A Guidebook, by Janet Ramage, 1997. Oxford University Press.
7. Solar Electricity, UNESCO Engineering Series, edited by Tomas Markvart, 1994. John Wiley & Sons.
8. Photovoltaics, by Robert Seippel, 1983. Publisher: Reston Publishing Company.
9. American Solar Energy Society www.ases.org
10. International Solar Energy Society www.ises.org
11. Solar Energy International (SEI) www.solarenergy.org
12. Latitude and Longitude www.artscipub.com/info/latlonofmajorcities.asp

Government Sites – General

1. US Department of Energy (DOE) www.energy.gov
2. US Department of the Interior www.doi.gov
3. US Bureau of Reclamation www.usbr.gov
4. US Department of Agriculture (USDA) www.usda.gov
5. Environmental Protection Agency (EPA) www.epa.gov
6. Food and Drug Administration (FDA) www.cfsan.fda.gov
7. National Institute for Occupational Safety and Health (NIOSH) www.cdc.gov/niosh
8. Mine Safety and Health Administration (MSHA) www.msha.gov
9. Federal Energy Regulatory Commission (FERC) www.ferc.gov
10. Nuclear Regulatory Commission (NRC) www.nrc.gov
11. National Climatic Data Center (NCDC) www.ncdc.noaa.gov

M. Department of Energy (DOE) Related Sites

1. Department of Energy (DOE) www.energy.gov
2. Energy Information Administration (EIA) www.eia.doe.gov
3. [Office of] Efficiency and Renewable Energy (EERE) www.eere.energy.gov
4. Office of Fossil Energy (in Dept of Energy) www.fossil.energy.gov
5. Electric Transmission and Distribution Office www.electricity.doe.gov
6. Science (Office of Science) www.sc.doe.gov
7. Nuclear Regulatory Commission (NRC) www.nrc.gov
8. Civilian Radioactive Waste Management (OCRWM) www.ocrwm.doe.gov
9. Yucca Mountain Project www.ocrwm.doe.gov/ymp/about/index.shtml
10. International Nuclear Safety Program http://insp.pnl.gov
11. International Nuclear Safety Center, Argonne Laboratory www.insc.anl.gov
12. National Energy Technology Laboratory (NETL) www.netl.doe.gov
13. National Renewable Energy Laboratory (NREL) www.nrel.gov
14. Oak Ridge National Laboratory www.ornl.gov
15. Los Alamos National Laboratory (LANL) www.lanl.gov/worldview
16. Pacific Northwest National Laboratory (PNL) www.pnl.gov
17. Starlight, from PNNL/DOE http://starlight.pnl.gov

Index

advanced solar cell technologies: 41–49

array (Solar Array):

    orientation and tilt: 33–40, 52–54

    overview: 15–18

    physical support: 30

    size of array: 23–28

basic operation of solar cells: 11–18

batteries and solar power: 22–23, 26–28

cooling of solar cells: 16, 44, 48

current and solar cells: 12–18

diode: 18

efficiency of solar power: 11, 13, 16, 21, 25, 41–49

electron–hole pair: 13, 17, 42

encapsulation: 29, 42–43

energy from sun: 24–25

equations for solar array: 24–25

heat and solar cells: 13, 16, 44–49

inverters: 15, 21–28

materials of solar cell:

    and advanced solar cell technologies: 42–48

    and energy to excite: 19–20

    and general process of solar cells: 12, 14, 16–17, 19

    specific materials used in solar cells: 20, 29, 45–46

n–type: 17–18

orientation of solar arrays: 16, 33–35

photovoltaic cells: 11–18

pn junction: 18

p–type: 17–18

PV: see "photovoltaic cells"

sizing an array: 23–28

solar collector: 11

solar concentrators: 46–48

temperature and solar cells: 16, 44, 46–48

thin films: 45–46

tilt angle (for solar array): 33–40, 52–54

wavelengths: 14, 19–20, 43–47, 51

www.ingramcontent.com/pod-product-compliance
Lightning Source LLC
Chambersburg PA
CBHW081242180526
45171CB00005B/514